Population and Settlement

● 1.1 Population dynamics

1 Look at Figure 2 on page 2 of the textbook. When did the world's population reach:

   a 1 billion? ....

   b 5 billion? ....

   c 7 billion? ....

2 When is the world's population forecast to reach 8 billion? ....

3 In your textbook, look at Table 1 on page 2, and read the paragraph on page 3 commenting on Table 1.

   a How many people were born in 2012? ....

   b How many people died in 2012? ....

   c By how much did the world's population increase in 2012? ....

4 When was the highest rate of global population growth? ....

5 List the five largest countries in terms of population in 2012.

   1 ....
   2 ....
   3 ....
   4 ....
   5 ....

6 Define the birth rate.

   ....

7 Which world regions have the highest and lowest birth rates?

   Highest ....
   Lowest   ....

8 If a country has a birth rate of 32/1000 and a death rate of 8/1000, what is its rate of natural increase?

   ....

Cambridge IGCSE Geography Workbook   Photocopying prohibited

1 POPULATION AND SETTLEMENT

9 Name the five stages of the model of demographic transition.

Stage 1 ....

Stage 2 ....

Stage 3 ....

Stage 4 ....

Stage 5 ....

10 Where and why in the model of demographic transition is the rate of population growth highest?

....

....

....

....

11 Define 'total fertility rate'.

....

....

12 Give one reason why the total fertility rate is a better/more detailed measure of fertility than the birth rate.

....

....

....

....

13 How can the infant mortality rate influence the level of fertility in a country?

....

....

....

....

14 a What type of graph is Figure 10 on page 8 of your textbook? ....

   b What name is given to the slanted line on the graph? ....

15 Describe the correlation between the total fertility rate and the percentage of girls enrolled in secondary education, illustrated by Figure 10.

....
....
....
....

16 Define 'life expectancy at birth'.

....
....
....
....

17 Which world regions have the highest and lowest life expectancy figures?

Highest ....

Lowest ....

18 In which type of country do infectious diseases such as malaria and tuberculosis kill many people?

....
....

19 Where in the world is HIV/AIDS most prevalent? ....

20 Describe two factors responsible for high rates of HIV/AIDS.

1 ....

2 ....

21 Define 'optimum population'.

....
....
....
....

22 Give two reasons why the world as a whole might be considered to be over-populated.

1 ....

....

2 ....

....

23 Define 'population policy'.

....

....

24 What is the name given to a population policy that:

a promotes large families? ....

b aims to reduce population growth? ....

25 How effective has China's one-child policy been in reducing fertility?

....

....

....

....

....

....

26 What problems has the one-child policy created in China?

....

....

....

....

....

....

27 Why has France, along with a number of other developed countries, taken measures to encourage fertility?

   ....

   ....

   ....

   ....

● 1.2 Migration

Look at Figure 1 on page 19 of the textbook.

1 Define 'migration'.

   ....

   ....

   ....

   ....

2 Explain the difference between push and pull factors in migration.

   ....

   ....

   ....

   ....

3 Give an example to explain voluntary migration.

   ....

   ....

   ....

   ....

4 Give an example to explain involuntary (forced) migration.

   ....

   ....

   ....

   ....

1 POPULATION AND SETTLEMENT

5 Look at Figure 4 on page 21 of your textbook. Why do more international migrants move from the South (the developing world) to the North (the developed world) rather than in the reverse direction?

....

....

....

....

6 Why is population movement within countries at a much higher level than movements between countries?

....

....

....

....

7 In which country is the largest ever internal migration currently taking place?

....

8 Define 'rural depopulation'.

....

....

9 Give two causes of rural depopulation.

1 ....

....

2 ....

....

10 Define 'counterurbanisation'.

....

....

1.2 Migration

11 Why has counterurbanisation occurred in so many developed countries?

....

....

....

....

12 Fill in two examples of the impact of international migration in each of the six cells in the table below.

Impact on countries of origin	Impact on countries of destination	Impact on migrants themselves
Positive 1 .... 2	1 .... 2	1 .... 2 ....
Negative 1 .... 2	1 .... 2	1 .... 2 ....

13 Define 'remittances'.

....

....

14 How high were global remittance flows in 2012? ....

15 Give two examples of the benefits of remittances to developing countries.

1 ....

2 ....

Cambridge IGCSE Geography Workbook   Photocopying prohibited

1.3 Population structure

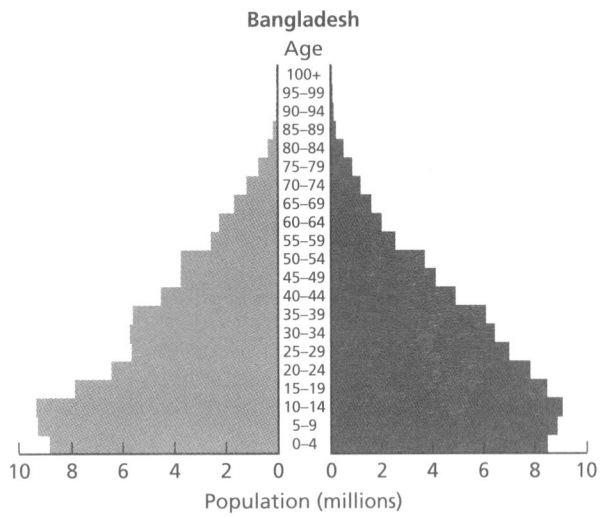

Bangladesh: the population pyramid for 2013

1 Define 'population structure'.

   ....

   ....

2 On the figure above:

   a add the labels 'females' and 'males' to the appropriate sides of the population pyramid.

   b draw horizontal lines to divide the pyramid into the young dependent population, the economically active population and the elderly dependent population.

3 How many years of age are represented by each bar on a population pyramid? ....

4 Suggest why the 0–4 and 5–9 bars on the population pyramid for Bangladesh are not as wide as the 10–14 bar.

   ....

   ....

5 The data in the figure above are shown in millions. What is the alternative way of showing data on a population pyramid?

   ....

6 Bangladesh is in which stage of demographic transition? ....

7 In which stages of demographic transition are:

   a Japan? ....

   b Niger? ....

   c the UK? ....

8 Why is the population structure of urban and rural areas in the same country sometimes markedly different?

   ....

   ....

   ....

   ....

   ....

   ....

9 Define the 'dependency ratio'.

   ....

   ....

10 What does a dependency ratio of 70 mean?

   ....

   ....

11 Is the dependency ratio generally higher in developed or developing countries?

   ....

12 Why is the dependency ratio an important factor for a country?

   ....

   ....

   ....

   ....

1 POPULATION AND SETTLEMENT

1.4 Population density and distribution

1 Explain the difference between population density and population distribution.

....

....

....

....

....

....

2 Explain the following terms:

a 'densely populated' ....

....

b 'sparsely populated' ....

....

3 Use your textbook to insert the population density figures for each world region in the table below:

Region	Population density (people per km²) in 2012
World	
More developed world	
Less developed world	
Africa	
Asia	
Latin America/Caribbean	
North America	
Europe	
Oceania	

Variations in world population density, 2012

4 How much higher is average population density in the less developed world compared with the more developed world? ....

5 Which world region has:

   a the highest population density? ....

   b the lowest population density? ....

6 Give three types of natural environment that are associated with very low population densities.

   1 ....
   2 ....
   3 ....

7 Discuss the factors that encourage high population density.

....
....
....
....
....
....
....
....

8 Complete the factfile below for the Canadian northlands, a sparsely populated region.

Location	
Population density	
Temperature	

1 POPULATION AND SETTLEMENT

Permafrost	
Economic activities	
Transport	
Settlement	

1.5 Settlements and service provision

1 Define the terms:

   a 'high-order services (or goods)' ....

   ....

   ....

   b 'low-order services (or goods)' ....

   ....

   ....

2 Arrange these settlements in descending order of size: *city*, *conurbation*, *hamlet*, *town*, *village*.

   ....

3 Study the map of the Lozère district in France below. Refer to page 40 of the textbook for data on the settlements.

   a i Describe the site of Mende.

   ....

   ....

   ii Describe the site of St-André-Capcèze.

   ....

   ....

1 POPULATION AND SETTLEMENT

iii Describe the situation of Bagnois-les-Bains.

....

....

b i Suggest why Mende has the highest population in the region.

....

....

ii Suggest why St-André-Capcèze has such a small population.

....

....

iii Identify, and justify, a likely location for another large settlement in the area.

....

....

iv Suggest why a large settlement might not develop in this location.

....

....

v Suggest two reasons why Badaroux has the second largest population in the study area.

....

....

....

....

● 1.6 Urban settlements

Urbanisation trends

	1950–2011	2011–2050
Asia	57.2	54.0
Latin America	14.0	32.5
Africa	6.8	13.2
Europe	9.0	6.8
North America	6.1	2.0
Oceania	0.6	0.5

World urbanisation (per cent of total urban increase)

1 Using the data in the table on page 14, draw two pie charts to show the changes in the percentage of total urban increase.

   a   1950—2011   b   2011—2050

2 In which region is the most urban increase predicted for 2011–2050?

   ....

3 In which region is the share of urban growth predicted to increase most between 1950–2011 and 2011–2050? ....

4 In which regions is the share of urban growth falling?

   ....

5 Briefly outline three advantages of living in an urban area.

   1 ....

   ....

   2 ....

   ....

   3 ....

   ....

6 Briefly outline three disadvantages of living in an urban area.

   1 ....

   ....

1 POPULATION AND SETTLEMENT

2 ....

....

3 ....

....

Cities by size and total population

The graph below shows total urban population by city size (millions) between 1970 and 2011.

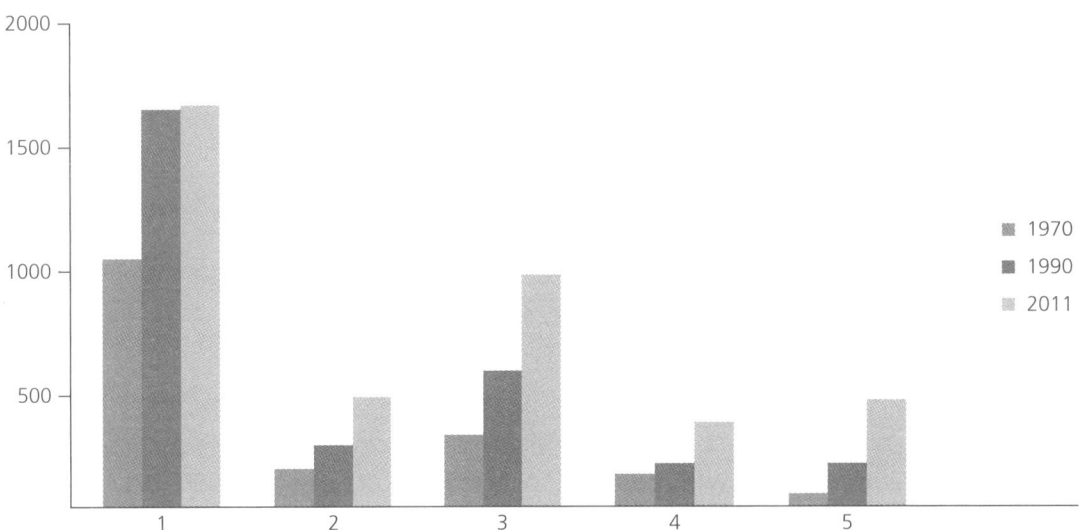

The projected data for 2025 are:

1	Less than 500 000	1966
2	500 000–999 999	516
3	1 000 000–4 999 999	1129
4	5 000 000–9 999 999	402
5	10 000 000 and over	630

1 Complete the graph above by adding the projected data for 2025.

2 a What is the name given to a city with over 1 million inhabitants? ....

b What is the name given to a city with over 10 million inhabitants? ....

3 Identify the size of city in which most people lived in 2011. ....

4 By which year were there more people living in cities of over 10 million compared with cities of

5 000 000–9 999 999? ....

5 Comment on the change in total population living in cities of different sizes between 2011 and 2025.

....

....

1.6 Urban settlements

....

....

The decline of the American shopping mall

Read the following extract about the decline of shopping malls in the USA and answer the questions below.

> The USA has about 1500 shopping malls. There are about 700 'super-regional' shopping centres and 800 smaller regional shopping malls. The super-regional 'megamalls' generally outperform their smaller counterparts. Between 15 and 50 per cent of shopping malls in the USA are projected to close in the coming decades. Recession, Americans increasingly shopping in central parts of the city, internet shopping and over-supply of shopping malls have contributed to their decline. Medium-sized shopping malls are set to fare worst. Shopping culture followed housing in the USA and moved to the suburbs following the Second World War. Northland Centre in Southfield, Michigan, was a shopping mall located just outside Detroit. It was built in 1954 but was sold to a developer in 2008.

1 Name a suburban shopping mall. ....

2 Explain why out-of-town centres (shopping malls) are in decline in the USA.

....

....

....

....

3 Outline the advantages of out-of-town shopping centres.

....

....

....

....

4 Briefly explain the disadvantages of out-of-town shopping centres.

....

....

....

Cambridge IGCSE Geography Workbook

1 POPULATION AND SETTLEMENT

Poverty and access to water in Detroit

Read the following extract about poverty and access to water in Detroit, USA, and answer the questions below.

> In July 2014 Detroit officials agreed to a pause in their attempt to recover $89 million in overdue water bills by cutting the supplies of people behind with their payments. In Detroit, the median household income is now less than half the national average. Anyone who owes $150 or is more than two months overdue with their water bills faces being cut off from the water supply. Some 80 000 customers are behind on their bills. In June 2014 the city council increased water prices by 8.7 per cent, which made Detroit's water bills nearly twice the US average.

1 Define 'median'.

   ....

   ....

2 a How does the median household income in Detroit compare with the US national average?

   ....

   ....

   b Comment on the household income in Detroit compared with the US national average.

   ....

   ....

   ....

   ....

3 How are the Detroit city authorities trying to raise their income?

   ....

   ....

4 How do water bills in Detroit compare with the rest of the USA?

   ....

   ....

5 Comment on the likely impacts and success of Detroit's water policy.

   ....

   ....

   ....

Urban quality of life

Study the data in the table below and answer the questions that follow.

City	Current population in city in millions (not metro region)	Central area density (population/km²)	Projected growth 2010–2025 (pp hour)	% of the country's population living in the city	GDP pp (US$)	Life expectancy (years)	% under 20	Car ownership rates per 1000	Daily water consumption (litres/pp)	Annual CO_2 emissions (kg pp)
Hong Kong	7.0	22 193	7	–	45 080	82.5	20.1	59	371	5 800
New York	8.1	15 353	9	2.8	55 693	77.6	25.7	209	607	7 396
Shanghai	15.5	23 227	26	1.0	8 237	81	16.0	73	493	10 680
London	7.6	8 326	1	12.4	60 831	79.2	23.8	345	324	5 599
Mexico City	8.6	12 860	10	8.4	18 321	75.9	32.9	360	343	5 862
Johannesburg	3.9	2 208	3	8.1	9 259	51	34.6	201	378	5 025
Mumbai	11.7	45 021	44	0.9	1 871	68.1	36.3	36	90	371
São Paulo	10.4	10 326	11	5.8	12 021	70.8	31.0	368	185	1 123
Istanbul	12.7	20 128	12	17.8	9 368	72.4	32.1	139	155	2 720

(Source: Based on *Urban Age*)

Note: pp = per person

1 Identify the city with the largest population. ....

2 Which city has the highest population density? ....

3 Identify the city that is projected to have the largest population growth between 2019 and 2025.

   ....

4 Which city accounts for the largest percentage of its country's population? ....

5 a Identify the richest city in terms of GDP per person. ....

   b Which is the poorest city in terms of GDP per person? ....

   c Calculate the ratio of the richest city to the poorest city, in terms of GDP per person. ....

6 In which city is life expectancy:

   a highest? ....

   b lowest? ....

7 Which city has the highest proportion of young people (i.e. those aged under 20 years)? ....

8 Comment on levels of car ownership among the cities.

   ....

   ....

1 POPULATION AND SETTLEMENT

9 Complete the graph below and comment on the relationship between GDP in urban areas and CO_2 emissions.

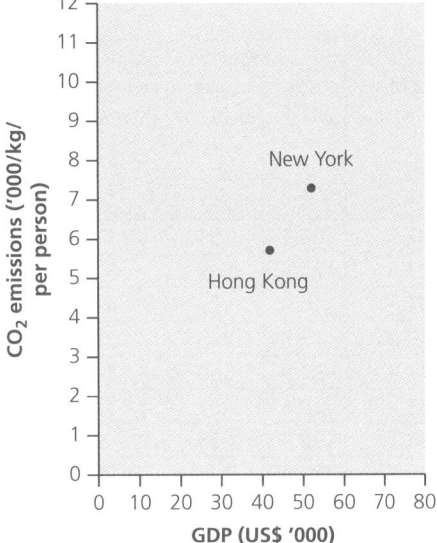

....
....
....
....
....
....
....

● 1.7 Urbanisation

The two photos below show a market town and a central business district.

Photo A

Photo B

1 Identify which photo shows:

   a a market town ....

   b part of a CBD. ....

2 Describe the main characteristics of Photo A.

....
....

3 Describe the main characteristics of Photo B.

....
....

2 The Natural Environment

2.1 Earthquakes and volcanoes

Tectonics

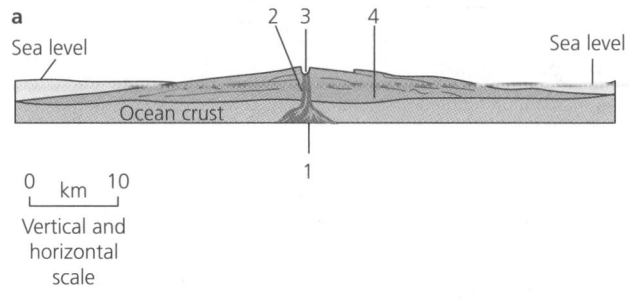

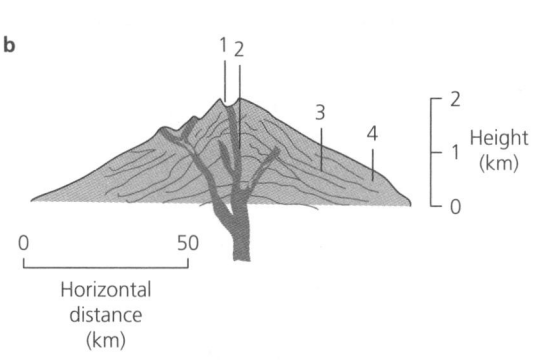

1 The figures on the left show a cone volcano and a shield volcano.

   a Identify both types of volcano.

   A ....

   B ....

   b For each volcano, identify features 1–4.

   A   B

   1 ....   1 ....

   2 ....   2 ....

   3 ....   3 ....

   4 ....   4 ....

   c Estimate the width and height of both types of volcano.

   A ....   B ....

   d Distinguish between magma and lava.

   ....

   ....

   ....

2 Describe the main characteristics of shield volcanoes and cone volcanoes.

   ....

   ....

   ....

   ....

Cambridge IGCSE Geography Workbook

The impact of volcanoes

Volcano	Year	Deaths	Major cause of deaths
Tambora, Indonesia	1815	92 000	Starvation
Krakatoa, Indonesia	1883	36 417	Tsunami
Mt Pelee, Martinique	1902	29 025	Ash flows
Nevado del Ruiz, Colombia	1985	25 000	Mudflows
Unzen, Japan	1792	14 300	Volcano collapse, tsunami
Laki, Iceland	1783	9 350	Starvation
Kelut, Indonesia	1919	5 110	Mudflows
Galunggung, Indonesia	1882	4 011	Mudflows
Vesuvius, Italy	1631	3 500	Mudflows, lava flows

Study the table above, which provides data on the most deadly volcanoes since 1600.

1 Which country has been most affected by deadly volcanoes? ....

2 Identify the most frequent hazard associated with volcanic eruptions.

....

3 Explain how volcanic eruptions may lead to starvation.

....

....

....

4 Distinguish between extinct, dormant and active volcanoes.

....

....

....

5 Suggest why the death toll from volcanoes is normally relatively low.

....

....

....

....

Earthquake impacts

The diagram below shows some of the characteristics of an earthquake and some impacts.

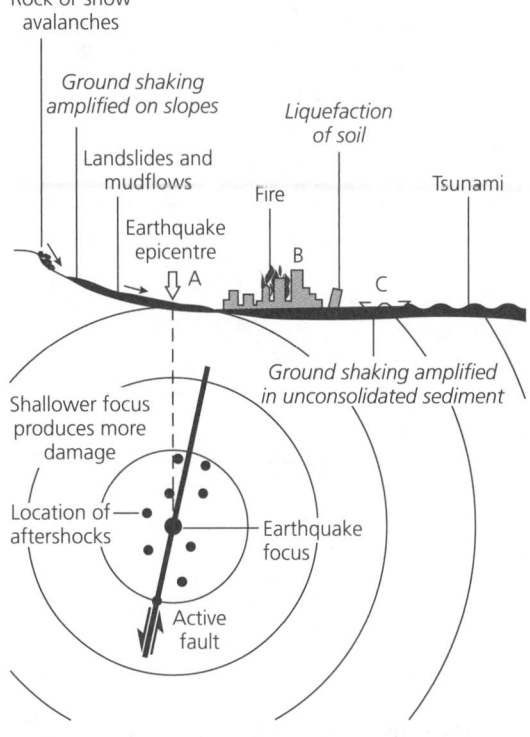

1 Define the terms 'epicentre' and 'focus' of an earthquake.

   ....

   ....

   ....

   ....

2 Explain why, for earthquakes of the same magnitude, shallow-focus earthquakes produce more damage than deep-focus earthquakes.

   ....

   ....

3 a Identify the location where damage from the earthquake is likely to be greatest, A, B or C.

   ....

   b Justify your choice.

   ....

   ....

   ....

   ....

2 THE NATURAL ENVIRONMENT

4 a Suggest what is meant by 'liquefaction of soil'.

....

b Why may liquefaction be a problem?

....

....

....

5 a What is an aftershock?

....

....

b In what ways may aftershocks be hazardous?

....

....

c Name one other hazard associated with earthquakes that is not shown on the diagram on page 23.

....

....

Living with volcanoes and earthquakes

1 Outline the ways of predicting volcanic eruptions.

....

....

....

....

2 Comment on the methods for predicting earthquakes.

....

....

....

....

....

3 Suggest ways in which volcanic eruptions can be managed.

....

....

....

....

4 Explain how earthquakes can be managed.

....

....

....

....

5 How does the management of volcanoes and earthquakes differ between MEDCs and LEDCs?

....

....

....

....

2.2 Rivers

Changes in sediment size

The diagram below shows changes in sediment size and sediment composition (make-up) in the Mississippi, USA.

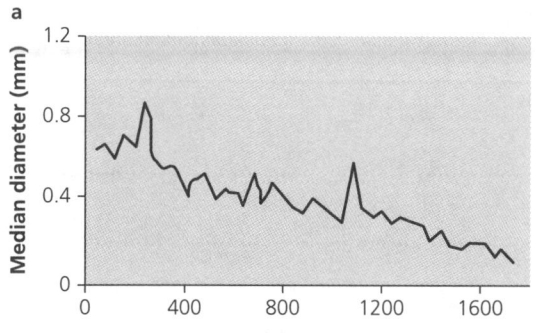
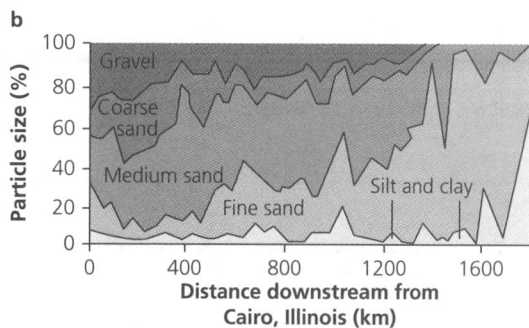

2 THE NATURAL ENVIRONMENT

1 Describe the changes in median sediment diameter with distance downstream.

　..

　..

　..

　..

2 a Estimate the percentage in each category sediment at 800 km.

　　Gravel　　　　............................%

　　Coarse sand　　............................%

　　Medium sand　　............................%

　　Fine sand　　　............................%

　　Silt and clay　　............................%

　b Describe the changes in the percentage of gravel downstream.

　　..

　　..

　c Describe the changes in the percentage of silt and clay downstream.

　　..

　　..

　　..

　　..

3 a How might the shape of particles be expected to change downstream?

　　..

　　..

　b Suggest why sediment shape may change downstream.

　　..

　　..

4 a Name one other type of load carried by a river. ....

　b Identify one rock type that may be a source of this load. ....

Waimakariri valley

Study the map of Arthur's Pass, New Zealand, on page 159 of the textbook.

1 a State the height of the highest point on the map.

   ....

   b Identify the height and location of the lowest point on the map.

   ....

2 a Describe the characteristics of the BB Trail.

   ....

   ....

   b Outline one hazard that may affect the tramping track.

   ....

   ....

3 a Describe three main characteristics of the Waimakariri river.

   ....

   ....

   ....

   b Describe three characteristics of the stream in square 7936.

   ....

   ....

   ....

   c Suggest reasons for the differences in the two rivers' characteristics.

   ....

   ....

   ....

Lowland rivers

The photograph below shows a river landscape in Eastern Europe. The white areas away from the river channel are covered in snow.

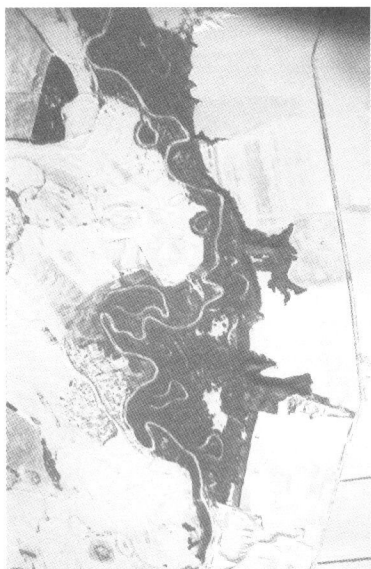

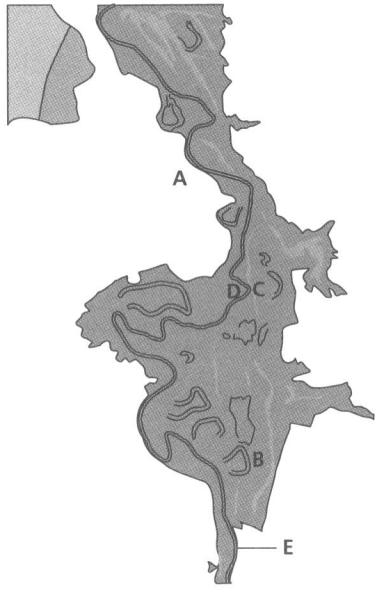

The sketch shows the same area.

1 a Identify the landforms A and B:

   A ....

   B ....

 b Identify the main process happening at C (outer bank) and D (inner bank):

   C ....

   D ....

 c Suggest a likely feature to be found at the edge of the river channel at E. ....

2 a With the use of a diagram(s) explain how oxbow lakes are formed.

b With the use of a diagram explain how feature E is formed.

Floods

1 Define 'flood'.

　...

　...

2 a Outline the disadvantages of floods.

　...

　...

　b Comment on the advantages that floods bring.

　...

　...

3 a Describe the natural causes of floods.

　...

　...

　b Outline the human factors that contribute to floods.

　...

　...

　...

　...

4 Explain why some floods are more intense than others.

　...

　...

　...

　...

　...

Floods

2.3 Coasts

Mapwork

1 Study Figure 6a and b on page 121 of the textbook, which shows the Cape Peninsula in South Africa.

   a State the highest point on the map. ....

   b Compare the relief of the slopes on the eastern side (Penguin Rocks) with those on the western side (Maclear Beach).

   ....

   ....

2 Study the map of Montego Bay, Jamaica, on page 42 of the textbook.

   a Identify the vegetation feature in square 5099 and 4800 (48100).

   ....

   ....

   b Suggest reasons for the location of the Yacht Club at 50102.

   ....

   ....

   ....

3 Study the map of south-west Tenerife on page 258 of the textbook, the note in the activities box on the top left of page 259, and the key below.

   Key

   ⌣ Swimming pool

   ▒ Bathing beach

   ⛴ Diving

   ⚓ Haven, ship landing

   🗼 Lighthouse

   a Name three beaches for bathing.

   ....

   ....

   ....

2 THE NATURAL ENVIRONMENT

b Where is it possible to go diving? ....

c Where is it possible to dock a boat? ....

d Which beach is easier to reach: Playa de Masca or Playa de Barranco Seco? Explain your answer.

....

....

....

Depositional features

1 Draw a labelled diagram to show how and why longshore drift occurs.

2 a Identify the landforms A, B and C on the diagram below.

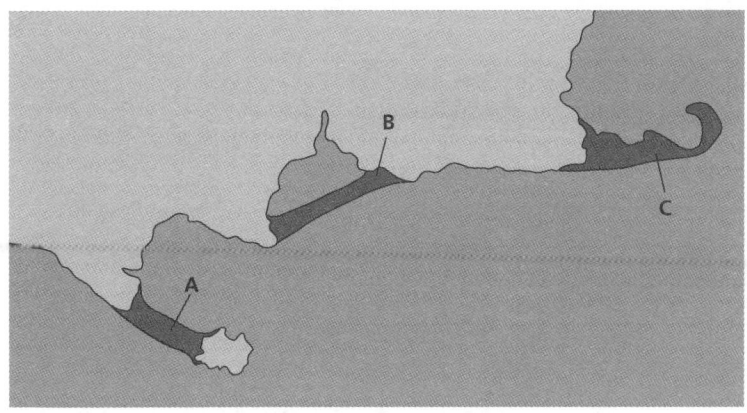

A ....

B ....

C ....

b Describe each of the three landforms.

A ....

....

B ....

....

C ....

....

3 Outline the conditions needed for the formation of landform C.

....

....

Coastal conflicts

The photograph below shows part of Barcelona port.

2 THE NATURAL ENVIRONMENT

The tables below provide data for the number of cruise passengers arriving at Barcelona port, and the amount of load from containers.

Year	No. of passengers to Barcelona/year
1980s	77 000
1998	466 000
2002	843 000
2005	1 250 000
2010	2 400 000
2013	2 600 000

Year	Container ship load (in '000 TEU – twenty foot equivalent)
1985	387
1995	683
2000	1 388
2005	2 100
2008	2 210
2010	1 931

(Source: Recent traffic dynamics in the European container port system, *Port Technology International*, Issue 58, 2013)

1 Complete the graph below showing the number of passengers and the amount of load arriving in Barcelona.

Key
— Container freight ('000 TEU) (twenty-foot equivalent)
---- No. of cruise ship passengers ('000)

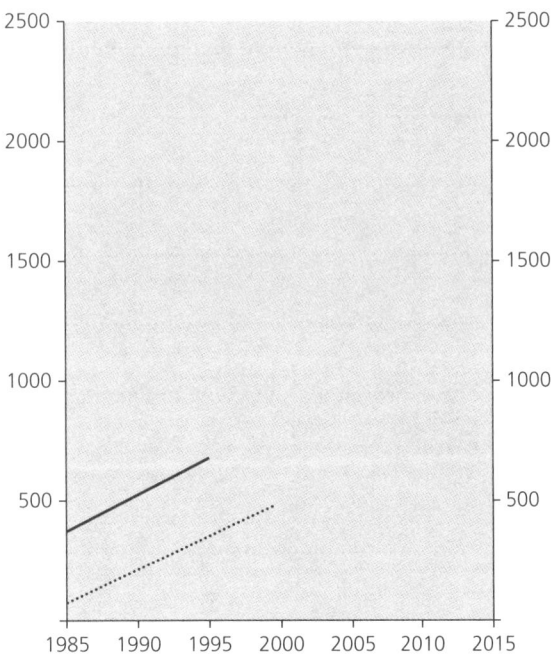

Graph to show changes in container trade and cruise ship passengers in Barcelona

2 Compare and contrast the changes in container load and number of cruise passengers.

....

....

....

....

3 Suggest some of the potential conflicts for users of the port of Barcelona, of the increase in the number of cruise passengers and container load. The photograph on page 33 may suggest some potential conflicts.

....
....
....
....
....
....

2.4 Weather

Clouds

1 Study the diagram below, which shows different cloud styles.

a Identify cloud types A, B, C and D.

A .... C ....

B .... D ....

b Describe the main characteristics of a stratus cloud and a cumulonimbus cloud.

....
....
....
....

c What are the highest clouds called? ....

d Which unit is used to measure cloud cover? ....

e Above which height are high clouds formed? ....

f At what height do low clouds generally have their base? ....

2 a What are the high clouds formed from?

....

b What are low clouds mainly formed from?

....

c What does the word nimbus refer to in cloud development?

....

● 2.5 Climate and natural vegetation

Microclimate changes around a school

The diagram below shows part of a school ground that was used for an investigation into 'variations in the school's microclimate'.

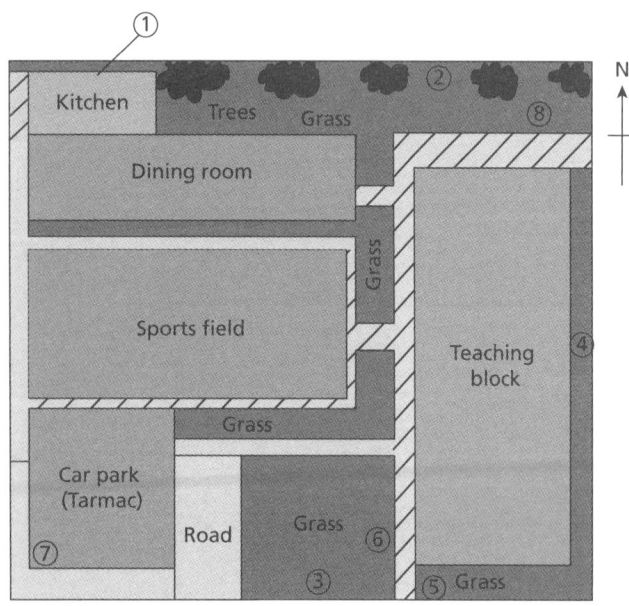

Suggest why:

1 Site 1 has the highest minimum temperature.

....

....

2.5 Climate and natural vegetation

2 Site 6 was warmer than Site 4.

   ....

   ....

3 Site 8 was warmer than Site 5.

   ....

   ....

4 Site 2 had the lowest recorded rainfall.

   ....

   ....

5 Site 7 was warmer than Site 3.

   ....

   ....

6 Sites 3 and 7 had the highest rainfall.

   ....

   ....

Tropical rainforests

Look at Figure 2 on page 150, which shows the distribution of areas of tropical rainforest.

1 Describe the distribution of tropical rainforests.

   ....

   ....

   ....

   ....

2 Study Figure 3 on page 150, which shows characteristics of the vegetation found in tropical rainforests.

   a What is the maximum height of the vegetation? ....

   b Describe the characteristics of the vegetation found at the top of the rainforest.

      ....

c Explain how this vegetation has adapted to conditions in the rainforest.

....

....

....

d Comment on the vegetation found at ground level in tropical rainforests.

....

....

....

Hot deserts

Study Figure 6 on page 152, which shows the distribution of areas of hot desert.

1 Describe the distribution of the world's hot deserts.

....

....

....

....

2 Describe the climate associated with hot deserts.

....

....

3 Outline the ways in which vegetation is adapted to conditions in hot deserts.

....

....

....

....

....

4 Outline the ways in which animals are adapted to conditions in hot deserts.

....

....

....

....

....

....

....

The effects of deforestation

The table below shows some of the uses of tropical rainforests.

Industrial uses	Ecological uses	Subsistent uses
• Charcoal • Medicines	• Soil erosion control • Flood control	• Fuelwood • Fodder for agriculture

1 Add 'Climate regulation', 'Weaving materials and dyes' and 'Tourism' to the above table.

2 Explain why deforestation may lead to increased flooding.

....

....

....

....

3 Explain why deforestation may lead to increased soil erosion.

....

....

....

....

2 THE NATURAL ENVIRONMENT

4 Explain how the composition of soil changes following deforestation.

....

....

....

....

5 How can deforestation lead to climate change?

....

....

....

....

Climate graphs

1 Using the data on page 147 (Manaus) and page 148 (Cairo) of the textbook, complete the following two climate graphs. The data for January to March have already been plotted.

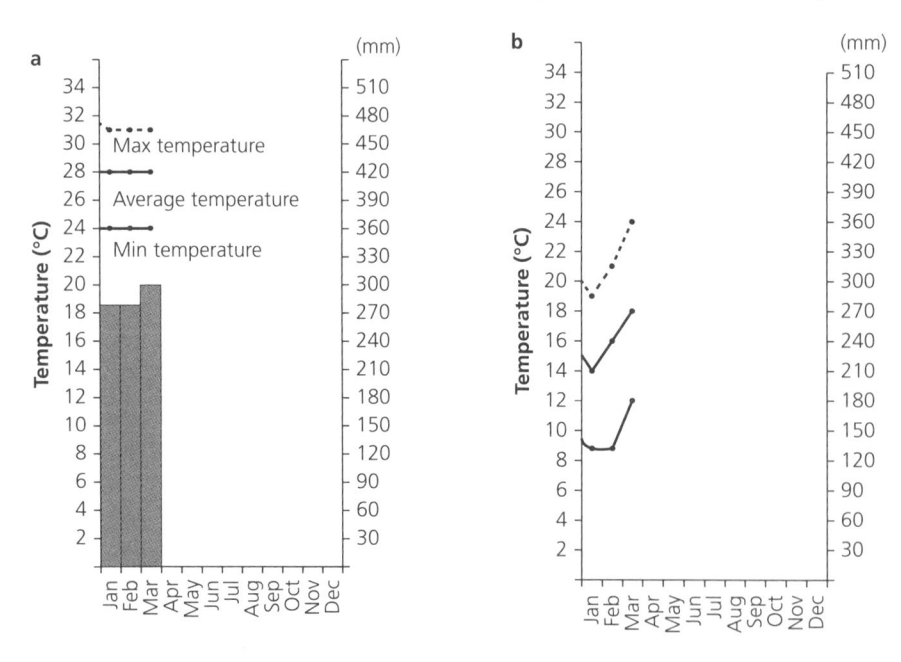

2 a State the range of average monthly temperatures in Manaus and Cairo.

Manaus ....

Cairo ....

b Identify the wettest month in Manaus. ....

c Which months have no rainfall in Cairo?

....

3 Compare the average number of sunshine hours in Manaus with Cairo.

   ....

   ....

4 Suggest why Manaus has a relatively low temperature range whereas Cairo has a high temperature range.

   ....

   ....

Weather maps

1 The diagram below shows the main weather symbols shown on weather maps.

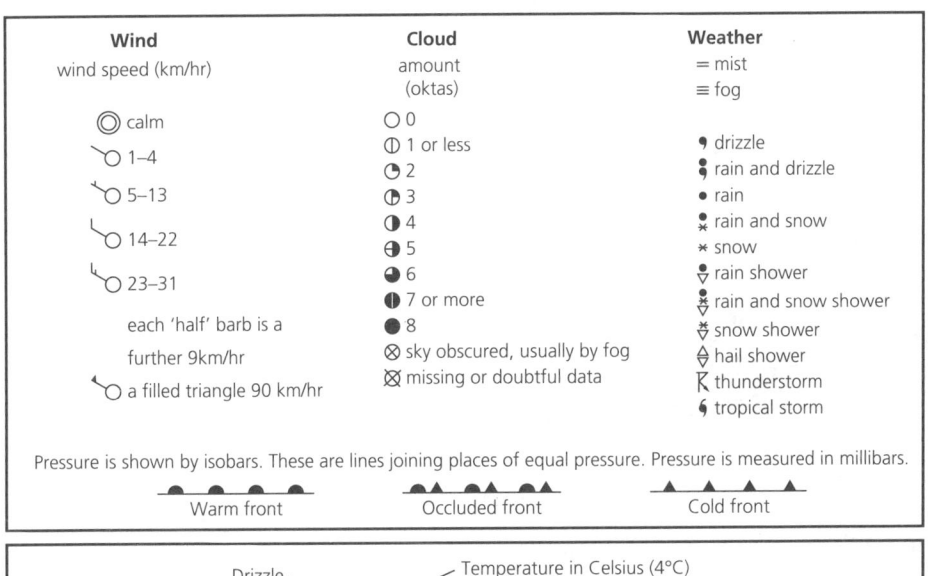

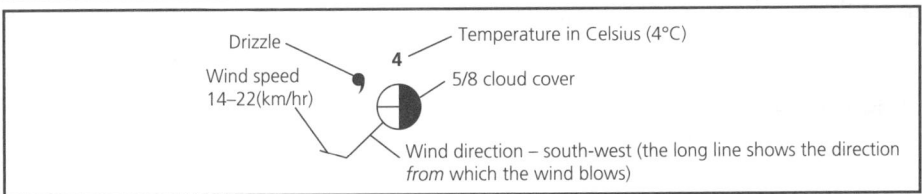

Wind	Cloud	Weather
wind speed (km/hr)	amount (oktas)	= mist
		≡ fog
◎ calm	○ 0	
○ 1–4	◐ 1 or less	ᛜ drizzle
○ 5–13	◐ 2	ᛜ rain and drizzle
○ 14–22	◐ 3	• rain
○ 23–31	◐ 4	✳ rain and snow
	◐ 5	✱ snow
each 'half' barb is a further 9km/hr	◐ 6	ᛜ rain shower
	◐ 7 or more	ᛜ rain and snow shower
	● 8	ᛜ snow shower
○ a filled triangle 90 km/hr	⊗ sky obscured, usually by fog	ᛜ hail shower
	⊠ missing or doubtful data	K thunderstorm
		ᛜ tropical storm

Pressure is shown by isobars. These are lines joining places of equal pressure. Pressure is measured in millibars.

⌒⌒⌒	⌒▲⌒▲	▲▲▲
Warm front	Occluded front	Cold front

Drizzle — Temperature in Celsius (4°C)
Wind speed 14–22(km/hr) — 4 — 5/8 cloud cover
Wind direction – south-west (the long line shows the direction *from* which the wind blows)

The map below shows the weather conditions for the British Isles during October.

2 THE NATURAL ENVIRONMENT

   a Identify the type of front located at B. ....

   b Identify the type of front nearest to point C. ....

2 Describe the weather conditions at:

   A ....

   B ....

   C ....

3 The following map shows a weather situation during August.

   a Identify the weather system at B. ....

   b State the air pressure in millibars. ....

4 a Describe the weather conditions at A (St Kitts and Nevis).

      ....

      ....

   b Describe the weather conditions at B.

      ....

      ....

   c Describe the weather conditions experienced in Florida, USA – location C.

      ....

      ....

3 Economic Development

3.1 Development

1 What is meant by the term 'development'?

....

....

....

....

2 Look at the photograph above and also at the photograph on the top left of page 162 in the textbook. Both photographs are from southern Mongolia. Describe what the photographs show and how they reflect the level of development in that region.

....

....

....

....

....

....

3 ECONOMIC DEVELOPMENT

3 Describe one numerical measure of the level of economic development in a country.

....

....

....

....

4 What is the 'development gap'?

....

....

....

....

5 Look at Figure 2 on page 163 of your textbook. Describe the global variation in GNP per capita in 2013.

....

....

....

....

6 Why are the following indicators considered to be good measures of development?

    a Literacy ....

....

....

    b Life expectancy ....

....

....

    c Infant mortality ....

....

....

....

7 List two other individual measures of development.

   1 ....

   2 ....

8 Why is the Human Development Index a better measure of development than the indicators considered in questions 6 and 7?

....

....

....

....

9 List the six countries with the highest level of human development in 2012.

   1 ....   4 ....

   2 ....   5 ....

   3 ....   6 ....

10 Insert the two labels 'Least developed countries' and 'Newly industrialised countries' into the correct places on the figure below:

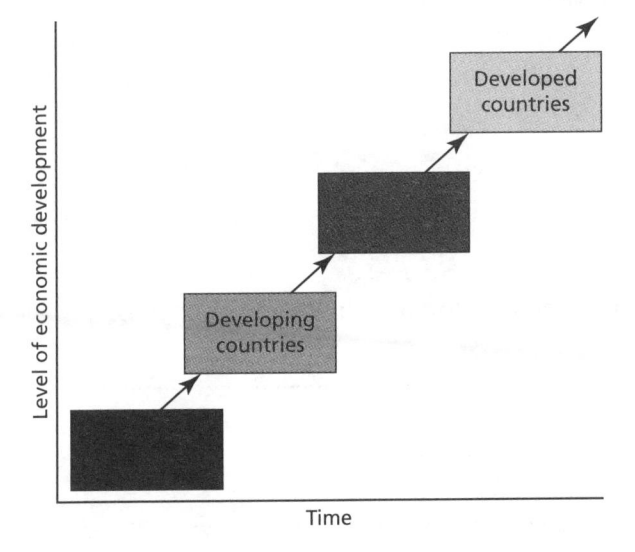

11 Where are the world's least developed countries located?

....

....

3 ECONOMIC DEVELOPMENT

....

....

12 Briefly state the reasons for the extremely low incomes of the least developed countries.

....

....

....

....

13 What is a newly industrialised country (NIC)?

....

....

14 Name the first four countries to be recognised as newly industrialised countries.

1 ....   3 ....

2 ....   4 ....

15 Name three other countries that have become NICs more recently.

1 ....

2 ....

3 ....

16 State two aspects of physical geography that have hindered development in various countries.

1 ....

....

2 ....

....

3.1 Development

17 Developing countries with good 'institutional quality' have been much more successful than those countries lacking this important development factor. What do you understand by the term 'institutional quality'?

....

....

....

....

18 Name a technique that is frequently used to show the extent of income inequality.

....

19 Look at Figure 9 on page 168 in the textbook. Name:

a four countries with very low income inequality

1 ....   3 ....

2 ....   4 ....

b four countries with very high income inequality.

1 ....   3 ....

2 ....   4 ....

20 Insert the two labels 'Economic core' and 'Periphery' in the correct places on the diagram below. Explain the meaning of each of these terms.

....

....

....

Cambridge IGCSE Geography Workbook

3   ECONOMIC DEVELOPMENT

21 What is the difference between 'regional economic divergence' and 'regional economic convergence'?

....

....

....

....

22 List four factors affecting inequalities within countries.

1 ....   3 ....

2 ....   4 ....

23 What is the difference between the formal sector and the informal sector within an economy?

....

....

....

....

....

....

24 Look at the photograph below.

Which sector of the economy does the photograph represent? Explain why.

....

....

25 Give two examples of jobs in each of the following:

The primary sector ....

....

The secondary sector ....

....

The tertiary sector ....

....

The quaternary sector ....

....

26 a Add the labels 'primary', 'secondary' and 'tertiary' in the correct places on the diagram.

b Explain how these three sectors change in importance over time as an economy develops.

....

....

....

....

....

....

27 Name a type of graph that can be used to show how different countries vary in employment structure.

....

Cambridge IGCSE Geography Workbook

3 ECONOMIC DEVELOPMENT

28 Define 'globalisation'.

....

....

....

....

29 What is a transnational corporation?

....

....

....

....

30 Why have advances in technology been important to the development of globalisation?

....

....

....

....

....

....

31 State and briefly explain three impacts of globalisation at the global scale.

1 ....

....

....

2 ....

....

....

Cambridge IGCSE Geography Workbook

3 ....
....
....
....

32 State two impacts of globalisation at the national scale.

1 ....
....
....

2 ....
....
....

33 Discuss one impossible impact of globalisation at the local scale.

....
....
....
....

● 3.2 Food production

1 Complete the blank spaces in the paragraph below.

Farming can be seen to operate as a .... with inputs,
.... and ....
The cultivation of crops is described as .... farming
while .... farming involves keeping livestock such as
.... and pigs. .... farming
involves cultivating crops and keeping livestock together.

2 What is the difference between subsistence farming and commercial farming?

....
....

3 ECONOMIC DEVELOPMENT

....

....

3 Complete the right-hand side of the table below to briefly describe the characteristics of the three types of farming.

Type of farming	Farming characteristics
Extensive farming	
Intensive farming	
Organic farming	

4 Give examples of extensive farming.

....

....

5 Briefly explain three physical factors that influence farming.

1 ....

3.2 Food production

....

....

2 ....

....

....

3 ....

....

....

6 Complete the blank spaces.

Economic factors influencing farming include markets, ... and .... Large farms allow .... of .... to operate which reduce the unit costs of production. Agricultural .... is the application of advanced techniques in farming. This is particularly important to the improvement of agriculture in developing countries. An important social/cultural factor in farming is land .... The influence of government on farming is classed as a .... factor. An example is the .... in the European Union.

7 What is irrigation?

....

....

8 Which is the most advanced type of irrigation – sub-surface, aerial or surface? ....

9 List four natural problems that can lead to food shortages.

1 ....   3 ....

2 ....   4 ....

10 Other economic and political factors that can also contribute to food shortages include:

1 ....

2 ....

3 ....

Cambridge IGCSE Geography Workbook   Photocopying prohibited

3 ECONOMIC DEVELOPMENT

11 Why is malnutrition such a significant problem in many poor countries?

....

....

....

....

....

12 Describe the three types of food aid.

....

....

....

....

....

13 Name the main organisations providing global food aid.

....

....

14 Give one criticism of the way food aid operates.

....

....

....

....

15 The Green Revolution has increased food production significantly in many developing countries.

   a State three advantages of the Green Revolution.

   ....

   ....

   ....

   ....

   ....

b State two disadvantages of the Green Revolution.

....

....

....

....

....

3.3 Industry

1 In the space provided, draw a simple diagram to show the three components of an industrial system.

2 Explain the difference between processing and assembly industries.

....

....

....

....

3 What are 'footloose industries'?

   ....

   ....

4 Is the iron and steel industry an example of heavy industry or light industry?

   ....

5 What is 'high-technology industry'?

   ....

   ....

   ....

   ....

   ....

6 Name two companies which manufacture high-technology products.

   ....

7 Where did high-technology industry first develop? ....

8 Why do high-technology industries often cluster together?

   ....

   ....

   ....

   ....

9 Give one example of a science park.

   ....

10 Explain two physical factors that influence the location of industry.

   1 ....

   ....

2 ....
....

11 Explain two human factors that influence industrial location.

1 ....
....

2 ....
....

12 What is meant by the term 'industrial agglomeration' and why does it occur?

....
....
....
....
....
....

13 Define an 'industrial estate'.

....
....

14 What are the reasons for grouping companies together on industrial estates?

....
....
....
....

15 How has the location of industry changed in recent decades:

a on a global scale? ....
....
....

b within individual countries? ....

....

....

c at an urban scale? ....

....

....

3.4 Tourism

1 Define 'tourism'.

....

....

2 a Look at Figure 1 on page 198 and describe the growth in global tourism from 1950 to 2010.

....

....

b What is the projected increase in global tourism between 2010 and 2020?

....

....

3 What were the reasons for the early development of tourism in the eighteenth and nineteenth centuries?

....

....

....

....

4 How many passengers did scheduled planes carry in:

   a 1970? ....   b 2012? ....

5 In the table below, insert three economic and three social factors that have influenced the growth of global tourism.

Economic factors	1 ....
	2 ....
	3 ....
Social factors	1 ....
	2 ....
	3 ....

6 Give one example of a political factor influencing tourism.

....

....

7 Define a tourist generating country.

....

....

8 Why is seasonality a major problem in many tourist destinations?

....

....

....

....

9 Use examples to distinguish between the direct and indirect economic impact of the tourist industry.

....

....

....

....

....

3 ECONOMIC DEVELOPMENT

10 Explain the meaning of 'economic leakages'.

....

....

....

....

....

....

11 State three economic benefits of tourism.

1 ....

....

2 ....

....

3 ....

....

12 In the table below, insert three positive and three negative social/cultural impacts of tourism.

Positive social and cultural impacts	Negative social and cultural impacts
1	1 ....
....	....
2	2 ....
....	....
3	3 ....
....	....

13 Define 'sustainable tourism'.

....

....

....

....

14 Suggest two ways that individual tourists can reduce their 'destination footprint'.

1 ....

....

2 ....

....

15 What is ecotourism?

....

....

● 3.5 Energy

1 State the non-renewable sources of energy.

....

....

2 Define 'renewable energy'.

....

....

....

3 What is the 'energy mix' of a country?

....

....

3 ECONOMIC DEVELOPMENT

4 Why is fuelwood such an important source of energy in the developing world?

....

....

....

....

5 In the table below identify four advantages and four disadvantages of nuclear power.

Advantages of nuclear power	Disadvantages of nuclear power
1	1 ....
2	2 ....
3	3 ....
4	4 ....

6 Why are most countries eager to develop renewable sources of energy?

....

....

....

....

3.5 Energy

7 a Why is hydro-electric power considered to be a traditional source of energy?

....

....

b Which four countries account for more than half of the world's HEP generation?

1 ....   3 ....

2 ....   4 ....

c Why is the opportunity for more large-scale HEP development very limited?

....

....

d Identify three problems associated with the development of HEP.

1 ....

2 ....

3 ....

8 Look at Figure 8 on page 212. Describe the increase in renewable energy consumption shown.

....

....

....

9 a State the worldwide capacity of wind energy.

....

b Which countries are the global leaders in wind energy?

....

c Why have so many countries invested in wind energy?

....

....

d Identify three concerns about the development of wind energy.

1 ....

2 ....

3 ....

Cambridge IGCSE Geography Workbook   Photocopying prohibited

3 ECONOMIC DEVELOPMENT

10 a What are biofuels?

....

....

b Which countries are the biggest producers of biofuels?

....

c What are the advantages of biofuels as stated by people that support their production?

....

....

....

....

d What are the disadvantages of biofuel production?

....

....

....

....

11 a Define 'geothermal energy'.

....

....

b Name the leading countries using geothermal electricity.

....

c Give three advantages of geothermal energy.

1 ....

2 ....

3 ....

d State three limitations of this form of energy.

1 ....

2 ....

3 ....

3.5 Energy

12 a What is the global capacity of solar electricity?

....

b Which countries are the leaders in the production of solar electricity?

....

c Describe the two ways in which solar electricity is produced.

1 ....

....

2 ....

....

d Discuss the advantages and disadvantages of solar electricity.

....

....

....

....

....

....

....

13 a How can electricity be produced from tides and waves?

....

....

....

....

b Why is so little electricity being currently produced by these methods?

....

....

....

....

Cambridge IGCSE Geography Workbook

3.6 Water

1 Why do water experts refer to a 'global water crisis'?

2 Define 'water supply'.

3 Why are dams and reservoirs so important to global water supply?

4 How are wells and boreholes operated to supply water?

5 a How does desalination work as a method of water supply?

   ....

   ....

   b What are the advantages and disadvantages of desalination?

   ....

   ....

   ....

   ....

6 Comment on two other methods of water supply which are not covered in the previous questions.

   1 ....

   ....

   ....

   2 ....

   ....

   ....

7 Look at Figure 4 on page 219. Describe and explain the differences in water use shown.

   ....

   ....

   ....

   ....

8 a Explain physical water scarcity.

   ....

   ....

   ....

   ....

3 ECONOMIC DEVELOPMENT

   b Explain economic water scarcity.

9 What is the difference between water stress and water scarcity?

10 Why do scientists expect water scarcity to become more severe in the future?

11 How can water management improve the water supply situation?

3.7 Environmental risks of economic development

1 Define 'pollution'.

   ....

   ....

   ....

2 What are the methods of human exposure to pollutants?

   ....

   ....

   ....

3 Which type of pollution has the most widespread effects on human health?

   ....

4 Name four major air pollutants.

   1 ....   3 ....

   2 ....   4 ....

5 Look at Figure 2 on page 227.

   a Define 'externalities'.

      ....

      ....

      ....

   b Explain the externality gradient and field shown in Figure 2.

      ....

      ....

      ....

      ....

3 ECONOMIC DEVELOPMENT

6 How serious is the global water pollution problem?

....

....

....

....

....

....

7 Name a major contributor to noise pollution. ....

8 a Define 'light pollution'.

....

....

b What are the causes and consequences of light pollution?

....

....

....

....

9 Explain the difference between incidental and sustained pollution.

....

....

....

....

10 Describe one major example of incidental pollution.

....

....

....

....

3.7 Environmental risks of economic development

....

....

11 Describe the natural greenhouse effect.

....

....

....

....

....

12 How has this effect been 'enhanced' by human activity?

....

....

....

....

....

13 List three of the main greenhouse gases.

1 ....

2 ....

3 ....

14 Discuss three consequences of enhanced global warming.

1 ....

....

....

2 ....

....

....

Cambridge IGCSE Geography Workbook

3 ....
....
....

15 Define 'soil erosion'.

....

....

16 How serious is the global soil erosion problem?

....

....

....

....

....

....

17 Explain the two major causes of soil degradation.

1 ....

....

2 ....

....

18 Define 'desertification'.

....

....

19 State two physical causes of desertification.

1 ....

2 ....

20 State two human causes of desertification.

   1 ....

   2 ....

21 Why is soil degradation a threat to food security?

   ....

   ....

   ....

   ....

   ....

   ....

22 Define:

   a 'resource management' ....

      ....

   b 'sustainable development' ....

      ....

23 Give one example of resource management.

   ....

24 Define:

   a 'recycling' ....

      ....

   b 're-use' ....

      ....

3 ECONOMIC DEVELOPMENT

25 What are the problems associated with landfill?

....

....

....

....

....

....

26 In terms of energy efficiency, define:

  a 'carbon credits' ....

   ....

  b 'community energy' ....

   ....

  c 'microgeneration' ....

   ....

27 Give three examples of the way that individuals can conserve energy.

  1 ....

   ....

  2 ....

   ....

  3 ....

   ....

4 Geographical Skills and Investigations

4.1 Geographical skills

Map skills

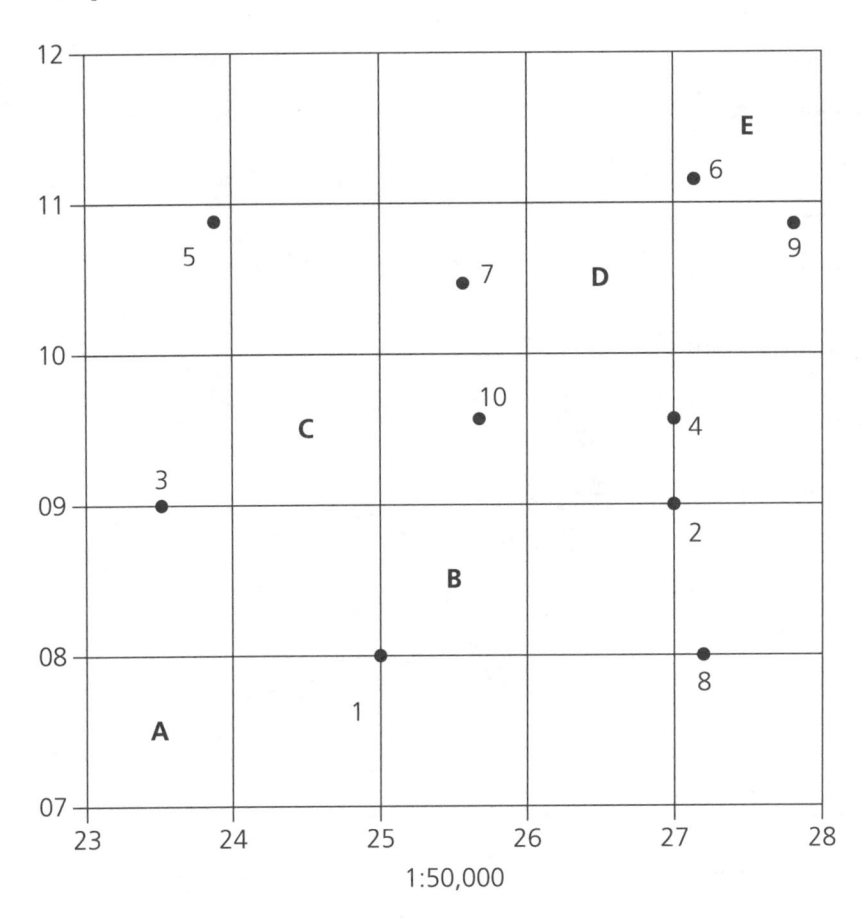

1 State the four-figure square reference for the following squares:

A ....

B ....

C ....

D ....

E ....

Cambridge IGCSE Geography Workbook

2 State the six-figure grid reference for the following points:

   1 ....
   2 ....
   3 ....
   4 ....
   5 ....
   6 ....
   7 ....
   8 ....
   9 ....
   10 ....

3 State the distance between the following points:

   2 to 4 ....

   1 to 8 ....

   3 to 2 ....

   4 to 9 ....

   7 to 2 ....

4 State the direction of:

   7 from 10 ....

   2 from 4 ....

   1 from 8 ....

   8 from 1 ....

   9 from 1 ....

The Publishers would like to thank the following for permission to reproduce copyright material:

Photo credits

Paul Guinness: p. 43, p. 48 *all*

Garrett Nagle: p. 20, p. 28, p. 33 *all*

Although every effort has been made to ensure that website addresses are correct at time of going to press, Hodder Education cannot be held responsible for the content of any website mentioned in this book. It is sometimes possible to find a relocated web page by typing in the address of the home page for a website in the URL window of your browser.

Hachette UK's policy is to use papers that are natural, renewable and recyclable products and made from wood grown in sustainable forests. The logging and manufacturing processes are expected to conform to the environmental regulations of the country of origin.

Orders: please contact Bookpoint Ltd, 130 Milton Park, Abingdon, Oxon OX14 4SB. Telephone: +44 (0)1235 827720. Fax: +44 (0)1235 400401. Lines are open from 9a.m.–5p.m. Monday to Saturday, with a 24-hour message answering service. You can also order through our website www.hoddereducation.com

© Paul Guinness and Garrett Nagle 2015

Hodder Education
An Hachette UK Company
Carmelite House, 50 Victoria Embankment, London EC4Y 0DZ

Impression number	5	4	3		
Year		2019	2018	2017	

All rights reserved. Apart from any use permitted under UK copyright law, no part of this publication may be reproduced or transmitted in any form or by any means, electronic or mechanical, including photocopying and recording, or held within any information storage and retrieval system, without permission in writing from the publisher or under licence from the Copyright Licensing Agency Limited. Further details of such licences (for reprographic reproduction) may be obtained from the Copyright Licensing Agency Limited, Saffron House, 6–10 Kirby Street, London EC1N 8TS.

Cover photo © arquiplay77 – Fotolia

Typeset in Frutiger LT Std 55 Roman 10/13 by Integra Software Services Pvt. Ltd., Pondicherry, India

Printed in the UK

A catalogue record for this title is available from the British Library

® IGCSE is the registered trademark of Cambridge International Examinations.

This text has not been through the Cambridge endorsement process.

ISBN 9781471845130

Cambridge IGCSE® Geography
Workbook

This workbook supports students using the Cambridge IGCSE Geography textbook, providing plenty of extra practice questions and activities.

- Perfect for using throughout the course – ensures students learn each topic thoroughly
- Covers the three geographical themes in the syllabus – Population and Settlement, the Natural Environment and Economic Development
- Answers available on the accompanying Teacher's Resource CD

Cambridge International Examinations and Hodder Education

Hodder Education works closely with Cambridge International Examinations and is an authorised publisher of endorsed textbooks for a wide range of Cambridge syllabuses and curriculum frameworks. Hodder Education resources, tried and tested over many years but updated regularly, are used with confidence worldwide by thousands of Cambridge students.

£7.00

ISBN 978-1-471-84513-0

HODDER EDUCATION
www.hoddereducation.com